IARC Diesel Exhaust and Lung Cancer

An Analysis

WRITTEN BY
Hank Campbell

A publication of the

AMERICAN COUNCIL
ON SCIENCE AND HEALTH

American Council on Science and Health
1995 Broadway, Suite 202
New York, New York 10023-5860
Tel. (212) 362-7044 • Fax (212) 362-4919
URL: http://www.acsh.org • Email: acsh@acsh.org

Publisher name: American Council on Science and Health
Title: IARC Diesel Exhaust and Lung Cancer: An Analysis
Author: Hank Campbell
Subject (general): Science and Health
Publication Year: 2015
Binding Type (i.e. perfect (soft) or hardcover): Perfect
ISBN: 978-0-9910055-8-1

Acknowledgements

The American Council on Science and Health appreciates the contributions of the reviewers named below:

W. Lawrence Beeson, Dr.P.H.
Professor, School of Public Health
Loma Linda University

Michael Berg, Ph.D.
Toxicologist
Center for Toxicology and Environmental Health, L.L.C.

William Bunn, M.D., J.D., M.P.H.
Adjunct Professor
Medical University of South Carolina
(and Northwesten and UIC)

James E. Enstrom, Ph.D., M.P.H.
University of California, Los Angeles

Tom Hesterberg, Ph.D., M.B.A.
Principal Toxicologist
Center for Toxicology and Environmental Health, L.L.C.

Rudolph Jaeger, Ph.D., DABT, ERT
President and Chief Scientist
CH Technologies (USA) Inc.

Table of Contents

The International Agency for Research on Cancer (IARC) was created in 1965 and operates under the auspices of the United Nations World Health Organization with the goal of examining evidence for carcinogenicity (involvement in causing cancer) of specific exposures. Their belief is that most cancers are linked to environmental factors and are therefore preventable by limiting those, whereas the weight of evidence has shown that most cancers are related to lifestyle (such as smoking or obesity) and genetics or other natural causes (e.g. random mutation.)

To quantify their levels of concern they use the following classifications:

- Group 1: Carcinogenic to humans
- Group 2A: Probably carcinogenic to humans
- Group 2B: Possibly carcinogenic to humans
- Group 3: Not classifiable as to carcinogenicity
- Group 4: Probably not carcinogenic

Because they are geared toward simpler assessments of hazard and not risk, their findings can be confusing to media and the public. For example, an October, 2015 monograph (volume 114) put processed meat in the same hazard category as cigarettes and plutonium, even though the actual risk of cancer from eating meat is minute compared to the risks of getting cancer from cigarettes.

Given the level of concern about diesel emissions due to a scandal resulting from discovery that the Volkswagen automobile manufacturer was manipulating software to achieve better performance in its diesel engine products, there has been some confusion among the public about the possible health effects of diesel emissions. While it is true that combustion engineers have long engaged in a trade-off between nitrogen-oxide emissions (NOx) and particulate matter in automobiles in order to optimize performance, the case of Volkswagen and IARC's findings on diesel emissions are not related.

There has long been concern that the hazard of particulate matter has been overstated and this document takes an unbiased look at the studies and methodology used.

IARC first examined epidemiological papers related to diesel exhaust (DE) in 1988 and classified it as probably carcinogenic to humans (Group 2A). In 2012 they evaluated it again (Benbrahim-Talla et al., 2012) and declared that it was a cause of lung cancer (sufficient evidence) and also claimed a positive association (limited evidence) with an increased risk of bladder cancer, which made it a Group 1 carcinogen, along with cigarettes, mustard gas and now bacon.

An unbiased examination of the methodology used in the meta-analysis, the template for choosing Working Group participants and the models in the studies underlying it show that their final document was far less definitive than it would need to be in order to justify it being classified in Group 1.

Because IARC is part of the World Health Organization, any re-classification has wide ranging implications, spanning from public health to legislative and ultimately financial standpoints. This is something they note with pride. As stated by IARC in the Preamble of their *Monographs on the Evaluation of Risk to Humans*,

> *"The monographs are used by national and international authorities to make risk assessments, formulate decisions concerning preventive measures, provide effective cancer control programmes, and decide among alternative options for public health decisions." (IARC, 2006)*

Yet even though they recognize how important their determinations are in policy, they often fail to note their own limitations or those of their monographs when speaking with popular media. Namely that the IARC *Monographs* program simply seeks to identify the causes of cancer, known as hazard identification, and does not make an attempt to define the degree to which each carcinogen presents a risk to public health. This very important distinction is often overlooked, especially when the results of IARC's classifications are communicated to members of the general public who may not have a full understating of the differences between hazard and risk.

IARC itself adds to this confusion by noting in the Preamble that the *Monographs* are an exercise in evaluation of cancer hazards, and then using the word 'risk' in the title.

To date, IARC has written 114 monographs and more than 400 agents have been identified as carcinogenic, probably carcinogenic, or possibly carcinogenic to humans. A breakdown by classification is reported in **Figure 1**.

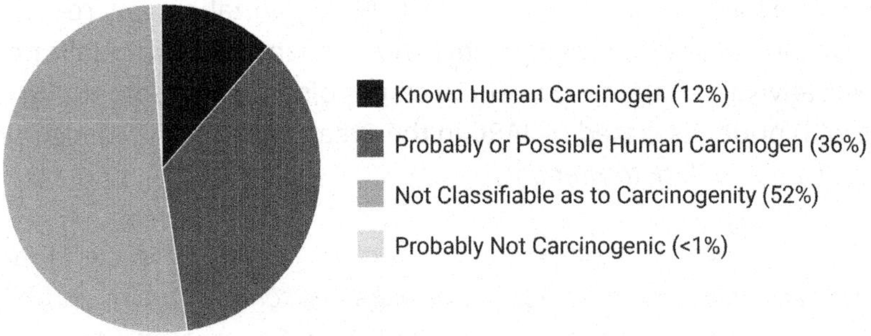

Figure 1: Breakdown of Evaluated Agents by Classification

Most recently, the outcomes of IARCs *Monograph* Program carcinogenicity assessments have received attention due to their assessment of high-priority chemicals or mixtures, including glyphosate, air pollution and red meat.

The purpose of this American Council on Science and Health White Paper is to provide commentary on IARC's 2012 assessment of diesel exhaust in the hopes that it will stimulate progressive thinking on the IARC decision-making process and also highlight some critical factors in the IARCs diesel exhaust assessment that should have warranted additional attention prior to an upgrade in diesel exhaust classification.

1
Selection of IARC Working Group

When a substance such as diesel exhaust is being evaluated for its potential to cause cancer, IARC staff identify a number of international experts, many with subject matter expertise to attend the *Monographs* meeting. These meeting participants are divided into 5 categories: the Working Group, Invited Specialists, Representatives of national and international health agencies, Observers with relevant scientific credentials, and the IARC Secretariat. Of these, only the Working Group members are allowed to vote in the final evaluation; however, members of the IARC Secretariat may also draft text or tables at the request of the Working Group or subgroup chair. As stated by IARC,

> *"Working Group Members generally have published significant research related to the carcinogenicity of the agents being reviewed, and IARC uses literature searches to identify most experts. Working Group Members are selected on the basis of (a) knowledge and experience, and (b) absence of real or apparent conflicts of interest." (IARC, 2013)*

Whereas the concept of Working Group selection appears to the casual observer as relatively straightforward in choosing the best experts who will be honored by an important task, that is not the case. IARC's Working Group selection process has instead become the subject of a number of critiques and responses published in the scientific literature in recent years (Erren, 2011; Wild and Straif, 2011; McLaughlin et al., 2011; McLaughlin et al., 2010; Wild and Cogliano, 2011; Gamble, 2012a; Pearce et al., 2015). The critiques, written in the form of editorials and letters to prominent scientific journals, largely revolve around the topic of conflict of interest or bias. When selecting the Working Group members, IARC screens for industry financial conflicts of interest and obligates those scientists with potential ties to industry to non-voting (and ultimately non-participating) meeting 'Observers'. Yet as was shown in the recent glyphosate program, a scholar who consults for environmental groups can meet their specified criteria even though such selection violates the spirit of the rule. A corporate consultant is no less or more conflicted than an environmental one.

There were also violations of IARC policy in this instance. Whereas the *Monographs* preamble states that the Observers may have an opportunity to speak at the meeting, this was not the case at the diesel exhaust monographs meeting, as Working Group members were strongly cautioned from interacting with the Observers. This is problematic since diesel companies are in the applied science and engineering fields, meaning the foremost scientists with expertise in diesel exhaust are often hired to consult within an industry setting or have been financially compensated for their expertise and, thus, were automatically prevented from serving as members of the Working Group due to a perceived conflict of interest. As noted, no similar prohibition is made against consultants for environmental groups.

It is a concern for the public that IARC, who negatively comment on the role of consultancy companies and industry (see Wild and Straif, 2011), does not want the best scientists, but instead chooses those who have never consulted. It also brings ethical worries, Wild and Straif note, because of the ethical implication hat because a scientist receives funding from industry they cannot be unbiased in reviewing data provided for IARC purposes – but consultants for other groups are considered without concern. Instead, one could make a similar argument for scientists in academia where a 'publish-or-perish' type attitude has become predominant as evidenced by the numerous journal article retractions and reports of 'pay-per-article' that have been recently noted (Kluger, 2015; Ferguson et al., 2014).

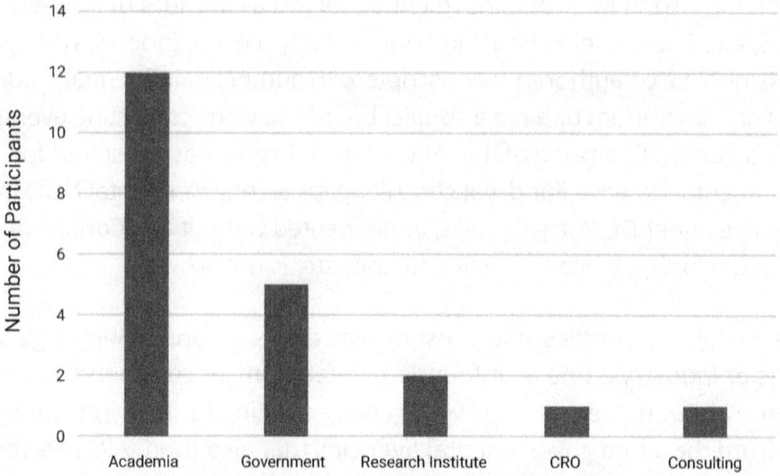

Figure 2: Breakdown of Diesel Exhaust Working Group

Some scientists have argued that by restricting conflict of interest to only industrial financial issues, *"IARC cannot see the elephant in the room"* - that IARC Working Groups are often composed of researchers who have a strong, lifelong interest in the compounds being evaluated (McLaughlin et al., 2011) much as if they had spent their whole career at a company. Because of their participant selection bias, it is not uncommon for manuscripts authored by IARC Working Group members to be discussed at the *Monograph* meeting or included into the final *Monograph* document. IARC claims that such career-interest scientists make up a minority of the Working Group because several agents are usually evaluated in a single, week-long meeting, and that *"care is taken to ensure that each study summary is written or reviewed by someone not associated with the study being considered"* (Pearce et al. 2015; IARC, 2006) but as noted by Gamble (2012a), it's an idealized belief, it is not required. Gamble provides an example of when the author of a study being reviewed for inclusion into the DE Monograph effectively argued against interpretative comments of his own work being included within the *Monograph*. A number of the

DE Working Group Members have either served as authors of influential studies, e.g. The Diesel Exhaust in Miners Study (DEMS) (doi: 10.1093/jnci/djs034), or have co-authored manuscripts with authors of influential studies. How is an honorarium or a fee a conflict but not obvious conflicting events? Dr. Chris Portier, Chairman of the IARC working Group, has consulted for Environmental Defense Fund, yet consulting for an environmental NGO is acceptable under Dr. Portier's rules, implemented without any Conflict of Interest disclosure, while consulting for industry is not allowed.

The belief that conflict of interest or bias exists in consultancy organizations or industry alone is unfounded. In fact, similar comments could be made about those scientists who receive funding for their research work from the same governmental agencies that also fund IARC. As very appropriately stated by Gamble (2012a),

> *"All persons have both recognized and un-recognized biases and vested interests. Potential conflicts of interest should be readily accessible for all to see, but opinions should be judged on factual accuracy and logic rather than source. Variable degrees of scientific honesty and independence are found in all affiliations and it cannot be assumed that they are only found in government, academia, and NGOs."*

In an effort to limit this conflict of interest, McLaughlin et al. (2010) proposes a Working Group made of experienced and well-trained investigators who are not professionally invested in the topic of the Working Group discussion. Meanwhile, Erren (2011) suggests a Working Group with an equal balance of both subject matter experts and non-experts, yet to-date IARC has refused to remove its selection bias.

The inclusion of opinions of subject matter experts – regardless of their employer, affiliations, or funding source – should be desired in such a scientific organization where decisions or conclusions have such wide ranging implications. Currently, the IARC process is closed to scientists and the public and only Working Group Members and Invited Specialists (and not others; such as Observers) are provided with draft documents prior to the meeting. When the final draft of the monograph is complete, it is not made available for anyone outside of IARC for review and comment prior to publication. A more transparent process is utilized by the United States Environmental Protection Agency (US EPA) when policy changes are proposed. The US EPA provides a draft document to anyone with an interest in the policy and a number of meetings that are open to the public are conducted to discuss other views on the proposed draft. Notably, the US EPA Health Assessment Document for Diesel Exhaust went through five drafts, each subjected to external review, prior to the final assessment being published (USEPA, 2002). Ultimately, such a document may take months, or even years to complete, but the final document is based on a more comprehensive knowledge of the science and is better accepted because the process was open and transparent.

2
Diesel Exhaust Monograph Meeting

In preparation for the Diesel Exhaust Monograph meeting, each of the Working Group members and Invited Specialists were presented with biological and epidemiological studies approximately 6 months prior to the IARC *Monographs* meeting in Lyon, France. This is done so that these meeting participants can prepare preliminary sections of the *Monographs* which are distributed to Working Group Members and other Invited Specialists for review prior to the meeting. As such, the first draft of the *Monographs* is often written by Working Group members in isolation of one another and prior to any critical debate or review with fellow scientists at the meeting. This is counterintuitive, scientists and the public expect that the *Monographs* would be drafted only after significant review and critique of the original and critical studies had taken place. It is an obvious flaw.

Another flaw is that the meeting in Lyon, France, lasted approximately one week. One week is a duration selected by IARC in an era when the available literature on a particular agent was typically much less voluminous than it is today (Gamble, 2012a). Because the duration of the meeting is so short – with respect to the amount of literature to examine – the time at the meeting is spent largely putting the finishing touches on different sections of the *Monograph* that were, as noted, pre-written in isolation prior to any critical debate. How daunting is that? The *Monograph* Volume 105 is over 700 pages long; meaning that to critically review the document in its entirety, the Working Group would have had to review over 100 pages (and check their accompanying references) per day. This is simply not feasible and it did not happen. For that reason, the allowance of additional time for review at the meeting would strengthen the IARC *Monograph* process. Another observer at the DE *Monograph* Meeting stated that *"out of his 24 days or so of IARC Meeting time, perhaps 3 hours was actually spent on robust scientific discourse"* (Gamble, 2012a).

3

Conflicts of Interest ignored when it comes to funding for the Monographs Programme

At the *Monographs* meeting, the publication (https://www.cdc.gov/niosh/topics/cancer/diesel/) of the Diesel Exhaust in Miners Study (DEMS) funded by the National Institute of Occupational Health and Safety (NIOSH) and the National Cancer Institute (NCI) was critical to IARCs reclassification of DE to a known human carcinogen. In fact, in 1998 the IARC Monograph Programme, which was largely funded by the NCI, recommended that diesel exhaust be listed as a high priority agent for re-evaluation because the NIOSH and NCI were conducting a large study of DE carcinogenicity in miners. The fact that the IARC decision relied heavily upon results from the DEM Study that was sponsored by the same agency that provides the primary source of funding for the *Monographs Programme* is a clear conflict of interest.

The final manuscripts from these studies were made available in March of 2012, only three months prior to the IARC *Monograph* meeting. It is also notable that one of the authors of this critical study, Roel Vermeulen, served on the IARC Working Group for Diesel Exhaust, and furthermore, as the Subgroup Chair of Exposure Data (Attfield et al., 2012; Silverman et al., 2012; IARC, 2013). As previously discussed, the idea that an author can critically evaluate his own work instead of simply defending the work that he has already published, while an industry consultant cannot objectively evaluate anything, is not reasonable and clearly introduces a source of bias into the IARC process. Since the initial publication of the DEM Study, several scientists have pointed out flaws in exposure assessment and/or analysis of results.

4

"Conflicted" status for those at the meeting who disagreed

At the DE meeting, those participants who disagreed with the DEM Study results and/or the lack of critical study review for the reasons noted were relegated to "conflicted" status, and their comments were largely ignored, more evidence that the group is not immune to emotional decision-making. It seems only logical that a publication or study with such wide-ranging implications be accepted and recognized by the scientific community as completely valid prior to being relied on as a primary factor by an international organization such as IARC, so penalizing those who noted the lack of critical review was not just petty, it led to bias in the commentary provided.

At the IARC Meeting, the DEM Study carried a lot of weight in moving DE to a Group 1 "Known Human Carcinogen" and notably superseded the sections of the Monograph Working Group that reviewed the animal studies. In this regard, IARC ultimately states that there is *sufficient evidence* in experimental animals for the carcinogenicity of whole diesel engine exhaust; however, it is fairly well documented (and even reported in the IARC DE *Monograph*) that laboratory animal studies are negative (e.g. not statistically significant) for tumors in all species except for rats. Even then, tumor formation in rats occurs only under lifetime DE exposure conditions with high exposure - durations and concentrations that lead to particle 'overload' in the lung (Hesterberg et al., 2012).

The particle 'overload' phenomenon is well documented in the scientific literature, and the US EPA has previously stated,

> *"Overload conditions are not expected to occur in humans as a result of environmental or most occupational exposures to DE. Thus, the rat lung tumor response is not considered relevant to an evaluation of the potential for a human environmental carcinogen." – USEPA HAD (2002)*

Nonetheless, while the IARC DE Working Group notes the species-specific nature of the particle 'overload' phenomenon with regard to rats, it is not discredited as a possible mechanism for DE induced carcinogenicity in humans because *"some aspect of the responses observed in rats are similar to those seen in humans exposed to diesel engine exhaust, which could help to elucidate the mechanism(s) of carcinogenic action in humans"* (IARC Monographs on the Evaluation of Carcinogenic Risks to Humans, No. 105.)

Nonetheless, despite the large availability of data which suggest that this is a species-specific response to extremely high concentrations of poorly soluble particles (not just DE), IARC still gives the results of these studies great weight in their evaluation of the carcinogenicity of DE in humans. This is similar to a previous IARC evaluation of carbon black where IARC stated that *"animal cancer data obtained under conditions of impaired lung clearance are relevant to humans"* (IARC, 2010).

It is worth mentioning that the DE particulate exposures which generate particle overload conditions in rats are 100s to 1000s of times greater than DE levels in the ambient environment or typical workplace.

It is also important to note that the DE used in the animal studies evaluated by IARC was generated by diesel engines manufactured before 2000. It is well documented (National Service Center for Environmental Publications, 2002) that the composition of DE is highly variable and among other things, highly dependent upon date of manufacture. For example, three general classes of diesel exhaust are often cited in the scientific literature: traditional DE (TDE; pre-1998), transitional DE, and new technology DE (NTDE; post-2007). In general, exhaust generated by NTDE engines contains orders of magnitude lower reductions of particulate matter (PM) and nitrogen oxides (NOx) when compared to TDE. Coinciding with this significant reduction in diesel PM is a substantial decrease in the amounts of specific particulate matter-associated compounds of toxicological concern (e.g. elemental carbon, organic carbon, polycyclic aromatic hydrocarbons (PAHs), nitro-PAHs, dioxins/furans, and metals). See the article by Tan, Pi-Qiang PQin Huanjing Kexue. 2013Mar;34(3):1150-5.

Because of this large difference in diesel exhaust composition in NTDE, many in the diesel industry requested that IARC delay the *Monograph* meeting until a chronic rodent inhalation study of NTDE could be completed. Doing so would allow for the results of such a study – which showed an absence of lung cancer in rats even after very high exposure concentrations for over two years – to be published and also allow an additional time period for the DEM Study to be reviewed and more formally accepted by the scientific community (HEI, 2015a; HEI, 2015b; HEI, 2015c). The results from this chronic NTDE carcinogenicity assessment were published by the Health Effects Institute (HEI) in January of 2015. While IARC noted that the human epidemiological studies were conducted before the modernization of diesel engine technology and that the use of *"NTDE will probably bring about an improvement with regard to public health"*, it is certain that the conclusions drawn in the *Monograph* are applicable to emissions from the modern diesel engine.

In this regard, one must question why IARC rushed to conduct their re-evaluation in 2012 instead of simply delaying it temporarily to address these issues. It is equivalent to doing a critical analysis of smart phones but only using models made before 2000.

5
Diesel Exhaust in Miners Study

IARC's re-evaluation of diesel exhaust relied heavily on the results of the DEM Study that, as previously noted, was only published just prior to the IARC DE *Monograph* meeting in 2012. Even further undermining its neutrality is that it was the anticipation of results from the DEM Study that was the driver for recommendation by the IARC *Monograph* Programme to re-evaluate DE as a carcinogen as early as 1998. Notably, the DEM Study is actually presented in a series of publications that culminates in two epidemiological papers; a case-control study and a cohort study (Attfield et al., 2012; Silverman et al., 2012).

Cohort studies and case control studies are observational studies, which means that the researchers observe the effect of a risk factor without trying to change who is exposed to it. A cohort study involves a group of people who are linked in some fashion, such as those born in a certain year or have various levels of exposure, while a case control study identifies people ("cases") who have an existing health problem and compares their prior exposure to the exposure of people without the health problem under investigation ("controls").. Cohort studies will be more relevant epidemiologically while case control studies are of less value because of confounders like recall bias.

The DEM Study likely represents one of the most important data sources for diesel exhaust epidemiological investigations due to its large variability of potential exposures, a large sample size (over 12,000 workers), and documentation of past lifestyle or occupational confounders (Gamble et al., 2012b). It evaluated the relationship between diesel exhaust exposure and lung cancer (among other cancer types) at a number of underground, non-metal mining facilities in the United States. These mines were selected because available records indicated low exposure to confounders (e.g. silica, radon, or asbestos), extensive diesel engine usage, and large numbers of workers combined with an extensive time history of diesel engine usage. However, despite the enormous dataset – the DEM Study took about 20 years to complete – there remain a number of uncertainties noted upon independent review, which, if valid, could limit the usefulness of the DEM Study results (Morfeld, 2012; Borak et al. 2011; Crump and van Landingham, 2012; Gamble et al., 2012b) and, due to its weight with the DE working group, the results of the analysis itself.

In the introduction to one of the DEM Study publications (Silverman et al., 2012), the authors highlight that two recent meta-analyses of epidemiological studies estimated the relative risk for lung cancer for those ever occupationally exposed to diesel exhaust to be 1.33 and 1.47, respectively, based on more than 35 studies (Bhatia et al, 1998; Lipsett and Campleman, 1999). Furthermore, they pointed out that a pooled analysis of 13,304 case subjects and 16,282 control subjects from 11 lung cancer case-control studies in Europe and Canada yielded an odds ratio (OR) of 1.31 for subjects in the highest vs lowest quartile of cumulative diesel exposure based on a job exposure matrix.

Results of this nature are consistent with that of past DE epidemiological investigations and suggest a relatively weak association between diesel exhaust exposure and lung cancer, an association that could simply be explained by confounding, such as cigarette smoking.

While the DEM Study attempted to address issues with confounders, a review of their initial findings brings to light several issues. For example, if diesel exhaust is a human carcinogen, why did the authors of the cohort mortality study find that those employed as surface-only workers who *"had very little to no contact with diesel equipment"* have a higher lung cancer mortality ratio than those working underground? This seems impossible considering that underground workers were exposed to, on average, 75 times greater diesel exhaust concentrations (as respirable elemental carbon; REC) than their above-ground counterparts. While the authors of the DEM Study suggest that this observed effect may be due to differences in smoking patterns, this is unlikely as the smoking prevalence (as reported in the companion study) in lung cancer cases was fairly similar for surface and underground workers.

Similarly, the results of the DEM Study suggest that exposure to high cumulative levels of DE (>304 $\mu g/m^3$-y) offers some protection from the carcinogenic effects of cigarette smoke for heavy smokers. Again, these results are counterintuitive to what one would expect toxicologically, and the authors suggest that enzymatic interactions or decreased DE particle deposition in the lungs of heavy smokers may support this observation. These additional questions should have been examined more closely prior to reliance on this study for an IARC evaluation.

At the forefront of debate on the DEM Study among diesel exhaust experts is the large degree of uncertainty surrounding the DE exposure assessment. As the exposure assessment forms the basis for any subsequent exposure-response evaluation, any variability or uncertainty within this component of the DEM Study could have greatly impacted the study outcome. The authors of the DEM Study continually state that they utilized *"historical quantitative DE exposure data"*; however, this statement is slightly misleading as their exposure assessment was built from both a combination of quantitative data (reflecting only a few years of exposure measurements) and numerous assumptions or estimations for many parameters such as feet horsepower and equipment inventories. Briefly, the DEM Study investigators selected respirable elemental carbon (REC) as an analytical surrogate for DE. However, data on REC were only primarily available for a four-year time period (1998-2001) captured as part of the DEM Study. To assess exposure prior to 1998, the investigators **estimated** REC concentrations using historic measurements of carbon monoxide (CO) collected back to 1976. Because CO data were limited in the mines prior to 1976 (diesel usage in underground mines began in 1947), the investigators used a 'double surrogate' approach and again **estimated** REC concentrations using **estimations** of CO based on historical information on equipment, equipment horsepower, hours of equipment used, and mine ventilation rates. It is notable that some of this information on equipment, equipment horsepower, hours of equipment used, and mine ventilation rates was **estimated** as well. Importantly, however, this information on equipment inventory and usage patterns was only available for a minority of years evaluated and, as stated by Vermeulen et al. (2010),

> *"the lack of inventories was compensated by a careful scrutiny of each mines production characteristics, trends of time in the number of diesel pieces used (for all the facilities, there was generally little change in equipment from year to year), and the number of years the equipment was used, as well as being supplemented by information from the interviews".*

While the DEM Study investigators believe that this approach is an improvement over earlier DE exposure evaluation procedures (Stewart et al.; 2011) it is unclear if an exposure assessment approach with such a degree of inherent uncertainty can be taken by the IARC DE Working Group as supporting a causal relationship between DE exposure and lung cancer.

Because of the rapid scheduling of the IARC Meeting so soon after the publication of the DEM Study, the DEM Study results did not undergo rigorous post-publication peer review by independent members of the scientific community. As stated by Gamble (2012a),

> *"It is troubling that despite 30+ years of research into diesels and cancer, two of the most influential papers were very recent and reviewers did not have time for adequate review (including replication of results by independent groups)".*

Since the publication of the DEM Study, a number of investigators have provided insightful comments on the DEM Study exposure evaluation process that, if addressed prior to the IARC DE *Monographs* Meeting, would have strengthened any conclusions drawn by the IARC Diesel

Exhaust Working Group (Borak et al., 2011; Gamble et al., 2012b; Crump and van Landingham, 2012; Morfeld, 2012). For example, it has been noted that (1) the colorimetric methods used to assess CO concentrations within the mines in the 1990s and early 2000s were imprecise and should not be relied upon for exposure assessment purposes aside from general, range-finding industrial hygiene uses; (2) the relationship between CO and REC in DE is variable and, in fact, out of the DE gases examined by the DEM Study authors, CO exhibited the weakest correlation with REC; (3) the historical relationship between CO and REC is not linear (Crump 2012) as suggested by the DEM Study authors; and (4) horsepower (and therefore fleet horsepower) is simply a poor predictor of DE emissions.

Lastly, one reviewer of the DEM Study noted that in the cohort mortality study, results from initial, or a *priori*, analyses did not reveal a clear relationship between lung cancer mortality and DE exposure. It was only after significant exploratory adjustments to the dataset (e.g. adjustments for worker location, 15-year exposure lag, exclusion of workers with less than 5 years tenure, and restricted exposures < 120 $\mu g/m^3$-y) was such a relationship observed (Gamble et al., 2012b). This raises the question as to whether or not several 'adjustments' to the data were made by the investigators until the results yielded the desired outcome, or being up front that some findings were a posteriori.

Given the weight of the task of reevaluating the carcinogenicity of DE, there should have been extreme caution on IARCs part to rely so heavily on the DEM Study which, at the time of the IARC DE *Monographs* Meeting, had yet to be validated by rigorous scrutiny in the scientific community. For example, since the IARC *Monograph* Meeting in 2012, the NCI and NIOSH have provided DEM Study data to outside scientists for review and assessment. The results of these analyses were published just recently with the authors reporting that the association between DE and lung cancer is less robust than what is reported by the DEM Study investigators (Crump et al. 2015; Moolgavkar et al., 2015). These authors first replicated the results of the DEM Study investigators and then explored the data in more depth focusing on alternative REC estimates of exposures (including one estimates of REC that lacked a reliance upon extrapolation from the CO data), confounding by radon (another known lung carcinogen), temporal factors in REC exposures, and age.

Both manuscripts highlight how simple uncertainties in data inputs can affect the resulting conclusions, and caution against the use of the DEM Study data for quantitative risk assessment (QRA) purposes without further evaluation of all important variables.

It is clear that such outside assessments should have been conducted prior to the results of the DEM Study being incorporated into the IARC evaluation. By not allowing time for such assessments to be completed and addressed by the DEM Study investigators, the results of IARCs DE assessment are confounded and will remain controversial.

6
Diesel Exhaust Summary and Conclusions

The question of diesel exhaust being a human carcinogen has long been a subject of debate among members of the scientific community. This is a topic of particular importance for the estimated 1.4 million workers in the United States and 3 million workers in Europe (Pronk et al. 2009) who are occupationally exposed to DE. IARC's reclassification of diesel exhaust from Group 2A *(probably carcinogenic to humans)* to Group 1 *(carcinogenic to humans)* make some sense given how the reassessment was conducted as explained above; however, significant questions remain as to the validity of that reassessment. As reported by Sun et al. (2014), who conducted a review of DE exposure and lung cancer risk for publications authored between 1970 and 2013, *"neither cohort nor case-control studies indicate a clear exposure-response relationship between DE exposure and lung cancer. Epidemiological studies published to date do not allow a valid quantification of the association between DE and lung cancer."* Sun et al. (2014) cite the lack of consistent (and objective) exposure information as being one of the main issues facing epidemiologists studying the potential health effects of diesel exhaust.

Nonetheless, a number of the IARC Working Group members continue to move forward with their DE assessments, going so far as to develop quantitative risk estimates where only one (presented by California Environmental Protection Agency Air Resources Board, now CARB, 1998) previously existed. It is notable that previous assessments by both the USEPA and HEI found the epidemiological data too uncertain to derive a quantitative risk estimate (USPEA, 2002; HEI, 1999). These newly developed estimates concluded that the number of lung cancer deaths from DE exposure ranged from 17-689 per 10,000 individuals in a 45-year employment occupational setting and 21 per 10,000 individuals in an average ambient setting of 0.8 µg/m^3 EC (Vermeulen et al., 2014).

Like the DEM Study, critical evaluation of these authors' results indicates methodological issues that limit their utility; insofar as to suggest that it would be inappropriate to base public policy upon the meta-analysis proposed by Vermuelen et al. (2014) (Crump, 2014; Morfeld and Spallek, 2015).

7

IARC: Shutting out complete scientific input

As currently stands, the IARC *Monograph* Programme relies heavily upon a process which lacks transparency and is closed to complete scientific input. Because such a process ultimately silences members of the scientific community that may have opposing views on the evaluation of data used to determine if a particular agent is carcinogenic, conclusions drawn at future IARC *Monograph* meetings will likely to continue to invite controversy. This is evident in IARC's recent classification of glyphosate (2015) and processed meat (Volume 114 of the IARC Monographs, The Lancet Oncology, Volume 16, No. 16, p1599–1600, doi:10.1016/S1470-2045(15)00444-1, 2015) not to mention cell phones (Volume 102 of the IARC Monographs, The Lancet Oncology Volume 12, No. 7, p624–626, July 2011, DOI: 10.1016/S1470-2045(11)70147-4). The importance of a balanced scientific assessment conducted in a time period sufficient to resolve important data-related issues cannot be understated. This is of particular importance with respect to diesel exhaust, since IARC's evaluation weighed so heavily on newly published epidemiological studies authored by (among others) members of the DE Working Group.

Such practices simply question the strength and/or validity of IARCs DE carcinogenicity assessment. As such, members of the scientific community and public policy decision makers should not look to IARC alone as the gatekeeper for carcinogenicity assessments; rather, IARC's evaluation should represent only one document in a compendium of assessments conducted by other credible organizations (US EPA, National Toxicology Program, or HEI) prior to establishing a causal link between an agent and a disease.

References

Attfield, M. D.; Schleiff, P. L.; Lubin, J. H.; Blair, A.; Stewart, P. A.; Vermeulen, R.; Coble, J. B., and Silverman, D. T. The Diesel Exhaust in Miners study: a cohort mortality study with emphasis on lung cancer. J Natl Cancer Inst. 2012 Jun 6; 104(11):869-83.

Benbrahim-Tallaa, L.; Baan, R. A.; Grosse, Y.; Lauby-Secretan, B.; El Ghissassi, F.; Bouvard, V.; Guha, N.; Loomis, D., and Straif, K. Carcinogenicity of diesel-engine and gasoline-engine exhausts and some nitroarenes. Lancet Oncol. 2012 Jul; 13(7):663-4.

Bhatia, R.; Lopipero, P., and Smith, A. H. Diesel exhaust exposure and lung cancer [see comments]. Comment in: Epidemiology 1998 Jan;9(1):4-6. Comment in: Epidemiology 1998 Jul;9(4):474. Epidemiology . 1998 Jan; 9(1):84-91.

Borak, J.; Bunn, W. B.; Chase, G. R.; Hall, T. A.; Head, H. J.; Hesterberg, T. W. ; Sirianni, G., and Slavin, T. J. Comments on the Diesel Exhaust in Miners Study. Ann Occup Hyg. 2011 Apr; 55(3):339-42; author reply 343-6.

California Environmental Protection Agency Air Resources Board, Proposed Identification of Diesel Exhaust as a Toxic Air Contaminant, June, 1998 https://www.arb.ca.gov/toxics/dieseltac/staffrpt.pdf

Crump, K. and Van Landingham, C. Evaluation of an exposure assessment used in epidemiological studies of diesel exhaust and lung cancer in underground mines. Crit Rev Toxicol. 2012 Aug; 42(7):599-612.

Crump, K. Meta-analysis of lung cancer risk from exposure to diesel exhaust: study limitations. Environ Health Perspect. 2014 Sep; 122(9):A230.

Crump, K. S.; Van Landingham, C.; Moolgavkar, S. H., and McClellan, R. Reanalysis of the DEMS Nested Case-Control Study of Lung Cancer and Diesel Exhaust: Suitability for Quantitative Risk Assessment . Risk Anal. 2015 Apr 10.

Environmental Protection Agency, Health Assessment Document for Diesel Engine Exhaust, May, 2002; https://nepis.epa.gov/EPA/html/DLwait.htm?url=/Exe/ZyPDF.cgi/300055PV.PDF?Dockey=300055PV.PDF

Erren, T. C. IARC's plea for traditional 'expert' working groups--a recipe for problems? Int J Epidemiol. 2011 Dec; 40(6):1727-8; author reply 1728-31.

Ferguson, C.; Marcus, A., and Oransky, I. Publishing: The peer-review scam. Nature. 2014 Nov 27; 515(7528):480-2.

Gamble, J. F. IARC evaluations of cancer hazards: Comment on the process with specific examples from volume 105 on diesel engine exhaust. Journal of Clinical Toxicology. 2012a; 2(6):e106.

Gamble, J. F.; Nicolich, M. J., and Boffetta, P. Lung cancer and diesel exhaust: an updated critical review of the occupational epidemiology literature. Crit Rev Toxicol. 2012b Aug; 42(7):549-98.

HEI. Diesel emissions and lung cancer: epidemiology and quantitative risk assessment. A special report of the Institute's Diesel Epidemiology Expert Panel. Cambridge, MA: Health Effects Institute; 1999 Jun.

HEI. Health Effects Institute Update. Summer 2015c. http://pubs.healtheffects.org/getfile.php?u=1105

HEI. Study of lifetime animal exposure to new technology diesel engine exhaust finds no lung cancer. Boston, MA: Health Effects Institute; 2015b Jan 27. http://pubs.healtheffects.org/getfile.php?u=1070

HEI. Synopsis of Research Report 184, Parts 1-4: Effects of Lifetime Exposure to Inhaled New-Technology Diesel Exhaust in Rats. Boston, MA:

Health Effects Institute; 2015a Jan. http://pubs.healtheffects.org/getfile.php?u=1068

Hesterberg, T. W.; Long, C. M.; Bunn, W. B.; Lapin, C. A.; McClellan, R. O., and Valberg, P. A. Health effects research and regulation of diesel exhaust: an historical overview focused on lung cancer risk. Inhalation Toxicology. 2012; 24(Suppl 1):1-45.

IARC. Carbon black. IARC Monogr Eval Carcinog Risks Hum. 2010; 93:43-191.

IARC. Diesel and gasoline engine exhausts. IARC Monogr Eval Carcinog Risks Hum. 2013; 105:39-486.

IARC. Glyphosate. IARC Monogr Eval Carcinog Risks Hum. 2015; 112:1-92.

IARC. Preamble to the IARC Monographs (amended January 2006). Lyon, France: International Agency for Research on Cancer; 2006. http://monographs.iarc.fr/ENG/Preamble/CurrentPreamble.pdf

Kluger, J. Modern Science has a Publish-or-Perish Problem. Time. 2015 Aug 20; http://omnifeed.com/article/time.com/4004391/modern-science-has-a-publish-or-perish-problem/

Lipsett, M. and Campleman, S. Occupational exposure to diesel exhaust and lung cancer: a meta-analysis. Am J Public Health. 1999 Jul; 89(7):1009-17.

McLaughlin, J. K.; Boffetta, P.; La Vecchia, C.; Lipworth, L.; Blot, W. J., and Tarone, R. E. Authors' Response: Problems with IARC's "expert" working groups. International Journal of Epidemiology. 2011; 40(6):1728-1729.

Mclaughlin, J. K.; Lipworth, L.; Tarone, R. E.; La Vecchia, C.; Blot, W. J., and Boffetta, P. Authors' Response: RE: A further plea for adherence to the principles underlying science in general and the epidemiologic enterprise in particular. International Journal of Epidemiology. 2010; 39(6):1679-1680.

Moolgavkar, S. H.; Chang, E. T.; Luebeck, G.; Lau, E. C.; Watson, H. N.; Crump, K. S.; Boffetta, P., and McClellan, R. Diesel Engine Exhaust and Lung Cancer Mortality: Time-Related Factors in Exposure and Risk. Risk Anal. 2015 Feb 13.

Morfeld, P. and Spallek, M. Diesel engine exhaust and lung cancer risks - evaluation of the meta-analysis by Vermeulen et al. 2014. J Occup Med Toxicol. 2015; 10:31.

Morfeld, P. Diesel exhaust in miners study: how to understand the findings? J Occup Med Toxicol. 2012 Jun 7; 7(1):10.

Pearce, N.; Blair, A.; Vineis, P.; Ahrens, W.; Andersen, A.; Anto, J. M.; Armstrong, B. K.; Baccarelli, A. A.; Beland, F. A.; Berrington, A.; Bertazzi, P. A.; Birnbaum, L. S.; Brownson, R. C.; Bucher, J. R.; Cantor, K. P.; Cardis, E.; Cherrie, J. W.; Christiani, D. C.; Cocco, P.; Coggon, D.; Comba, P.; Demers, P. A.; Dement, J. M.; Douwes, J.; Eisen, E. A.; Engel, L. S.; Fenske, R. A.; Fleming, L. E.; Fletcher, T.; Fontham, E.; Forastiere, F.; Frentzel-Beyme, R.; Fritschi, L.; Gerin, M.; Goldberg, M.; Grandjean, P.; Grimsrud, T. K.; Gustavsson, P.; Haines, A. ; Hartge, P.; Hansen, J.; Hauptmann, M.; Heederik, D.; Hemminki, K.; Hemon, D.; Hertz-Picciotto, I.; Hoppin, J. A.; Huff, J.; Jarvholm, B.; Kang, D.; Karagas, M. R.; Kjaerheim, K.; Kjuus, H.; Kogevinas, M.; Kriebel, D.; Kristensen, P.; Kromhout, H.; Laden, F.; Lebailly, P.; LeMasters, G.; Lubin, J. H.; Lynch, C. F.; Lynge, E.; 't Mannetje, A.; McMichael, A. J.; McLaughlin, J. R.; Marrett, L.; Martuzzi, M.; Merchant, J. A.; Merler, E.; Merletti, F.; Miller, A.; Mirer, F. E.; Monson, R.; Nordby, K. C.; Olshan, A. F.; Parent, M. E.; Perera, F. P.; Perry, M. J.; Pesatori, A. C.; Pirastu, R.; Porta, M.; Pukkala, E.; Rice, C.; Richardson, D. B.; Ritter, L.; Ritz, B.; Ronckers, C. M.; Rushton, L.; Rusiecki, J. A.; Rusyn, I.; Samet, J. M.; Sandler, D. P.; de Sanjose, S.; Schernhammer, E.; Costantini, A. S.; Seixas, N.; Shy, C.; Siemiatycki, J.; Silverman, D. T.; Simonato, L.; Smith,

A. H.; Smith, M. T.; Spinelli, J. J.; Spitz, M. R.; Stallones, L.; Stayner, L. T.; Steenland, K.; Stenzel, M.; Stewart, B. W.; Stewart, P. A.; Symanski, E.; Terracini, B.; Tolbert, P. E.; Vainio, H.; Vena, J.; Vermeulen, R.; Victora, C. G.; Ward, E. M.; Weinberg, C. R.; Weisenburger, D.; Wesseling, C.; Weiderpass, E., and Zahm, S. H. IARC monographs: 40 years of evaluating carcinogenic hazards to humans. Environ Health Perspect. 2015 Jun; 123(6):507-14.

Pronk, Anjoeka; Coble, Joseph; and Stewart, Patricia; Occupational exposure to diesel engine exhaust: A literature review. J Expo Sci Environ Epidemiol. 2009 Jul; 19(5): 443–457

Silverman, D. T.; Samanic, C. M.; Lubin, J. H.; Blair, A. E.; Stewart, P. A.; Vermeulen, R.; Coble, J. B.; Rothman, N.; Schleiff, P. L.; Travis, W. D.; Ziegler, R. G.; Wacholder, S., and Attfield, M. D. The Diesel Exhaust in Miners study: a nested case-control study of lung cancer and diesel exhaust. J Natl Cancer Inst. 2012 Jun 6; 104(11):855-68.

Stewart, P. A.; Coble, J. B.; Vermeulen, R.; Blair, A.; Lubin, J.; Attfield, M., and Silverman, D. T. Reply to Borak et al. Comments on the Diesel Exhaust in Miners Study. Annals of Occupational Hygiene. 2011; 55(3):343-346.

Sun, Y.; Bochmann, F.; Nold, A., and Mattenklott, M. Diesel exhaust exposure and the risk of lung cancer-a review of the epidemiological evidence. Int J Environ Res Public Health. 2014; 11(2):1312-40.

. Tan, Pi-Qiang PQ. Emission characteristics of polycyclic aromatic hydrocarbons in exhaust particles from a diesel car (translated from Chinese) Huanjing kexue 2013 Mar;34(3):1150-5

USEPA. Health assessment document for diesel engine exhaust. Washington, DC: National Center for Environmental Assessment, U.S. Environmental Protection Agency; 2002 May; EPA/600/8-90/057F.

Vermeulen, R.; Coble, J. B.; Lubin, J. H.; Portengen, L.; Blair, A.; Attfield, M. D.; Silverman, D. T., and Stewart, P. A. The Diesel Exhaust in Miners Study: IV. Estimating historical exposures to diesel exhaust in underground non-metal mining facilities. Ann Occup Hyg. 2010 Oct; 54(7):774-88.

Vermeulen, R.; Portengen, L.; Silverman, D. T.; Garshick, E., and Steenland, K. Meta-analysis of lung cancer risk from exposure to diesel exhaust: Vermeulen et al respond. Environ Health Perspect. 2014 Sep; 122(9):A230-1.

Wild, C. P and Straif, K. Author's Response: Expert Working Groups--a reliable recipe. International Journal of Epidemiology. 2011; 40(6):1730-1731.

Wild, C. P. and Cogliano, V. J. A plea on behalf of expert evaluation and the experts involved. Int J Epidemiol. 2011 Feb; 40(1):253-61.

BOARD OF SCIENTIFIC AND POLICY ADVISORS

BOARD OF SCIENTIFIC AND POLICY ADVISORS (CONTINUED)

BOARD OF SCIENTIFIC AND POLICY ADVISORS (CONTINUED)

BOARD OF SCIENTIFIC AND POLICY ADVISORS (CONTINUED)

Made in the USA
Monee, IL
04 November 2026

67374703R00030